ANIMALS OF THE GRASSLANDS

HUNGRY HYENAS

THERESA EMMINIZER

PowerKiDS press.
New York

Published in 2021 by The Rosen Publishing Group, Inc.
29 East 21st Street, New York, NY 10010

First Edition

Portions of this work were originally authored by Ethan Grucella and published as *Hyenas*. All new material in this edition authored by Theresa Emminizer.

Editor: Theresa Emminizer
Book Design: Michael Flynn

Photo Credits: Cover, p. 1 gualtiero boffi/Shutterstock.com; series background Chantal de Bruijne/Shutterstock .com; p. 5 (brown hyena) EcoPrint/Shutterstock.com; p. 5 (striped hyena) J. Natayo/Shutterstock.com; p. 5 (spotted hyena) Federica Cordero/Shutterstock.com; p. 7 PickledImages/iStock/Getty Images; p. 9 Picture by Tambako the Jaguar/Moment/Getty Images; p. 11 MicheleB/Shutterstock.com; p. 13 W. Wisniewski/Corbis NX/Getty Images; p. 15 Jane Rix/Shutterstock.com; p. 17 Heinrich van den Berg/Oxford Scientific/Getty Images; p. 19 Nomads Nature/500px Plus/Getty Images; p. 21 Deon De Villiers/500px Prime/Getty Images; p. 22 Krakenimages.com/Shutterstock.com.

Library of Congress Cataloging-in-Publication Data

Names: Emminizer, Theresa, author.
Title: Hungry hyenas / Theresa Emminizer.
Description: New York : PowerKids Press, [2021] | Series: Animals of the grasslands | Includes index.
Identifiers: LCCN 2020001359 | ISBN 9781725317024 (paperback) | ISBN 9781725317048 (library binding) | ISBN 9781725317031 (6 pack)
Subjects: LCSH: Hyenas--Africa--Juvenile literature.
Classification: LCC QL737.C24 E46 2021 | DDC 599.74/3096--dc23
LC record available at https://lccn.loc.gov/2020001359

Manufactured in the United States of America

CPSIA Compliance Information: Batch #CSPK20. For Further Information contact Rosen Publishing, New York, New York at 1-800-237-9932.

CONTENTS

Hello, Hyenas!

There are three species, or kinds, of hyenas—brown, spotted, and striped. In this book, we'll look at the spotted hyena. Spotted hyenas are the biggest of the three species. Hyenas can be found across Africa in many different **habitats**, including **grasslands**.

brown hyena

striped hyena

spotted hyena

Hyenas have big heads, thick necks, strong jaws, and rounded ears. Their front legs are longer than their back legs. No two hyenas have the same markings. Hyenas might look a bit like dogs, but they're actually more like cats.

Laughing Loudly

Hyenas are noisy animals. They **communicate** with each other by howling, growling, and more. Spotted hyenas are sometimes called laughing hyenas. That's because their loud calls sound like people laughing! A hyena's "laugh" can be heard up to 3 miles (5 km) away.

Part of the Clan

Spotted hyenas live in groups called clans. Female, or girl, hyenas lead clans. A clan can have as many as 100 members. They work together to hunt and raise cubs. Each clan has its own **territory**, with a home base called a den.

Hyena Cubs

Female hyenas can weigh up to 190 pounds (86 kg). They give birth to one or two cubs a year. Cubs nurse from their mothers and live in the den until they're about a year old. An adult hyena "babysits" the cubs.

Built to Hunt

Hyenas are powerful hunters. They work together to bring down prey, which can be anything from rhinos to rabbits. They have good hearing and sharp eyesight, even in the dark. They are fast, running up to 37 miles (60 km) per hour.

They'll Eat Anything

Hyenas eat almost every part of an animal, including skin, bones, and teeth! They can eat a third of their body weight in just one meal. Hyenas are known for being **scavengers**, but actually hunt most of their food.

Hyenas in Trouble

People are the biggest threat, or danger, to hyenas. As human populations grow, more people are moving into hyena habitats. Hyenas are seen as pests and are often hunted and killed. They're often blamed for killing farm animals and stealing food.

We Need Hyenas!

Many people see hyenas as unwanted pests. With their odd looks, loud "laughing" calls, and scavenging ways, they've often been misunderstood. But hyenas are important members of the grassland **ecosystem**. Helping people understand hyenas' value will be key to their future.

Fancy Features

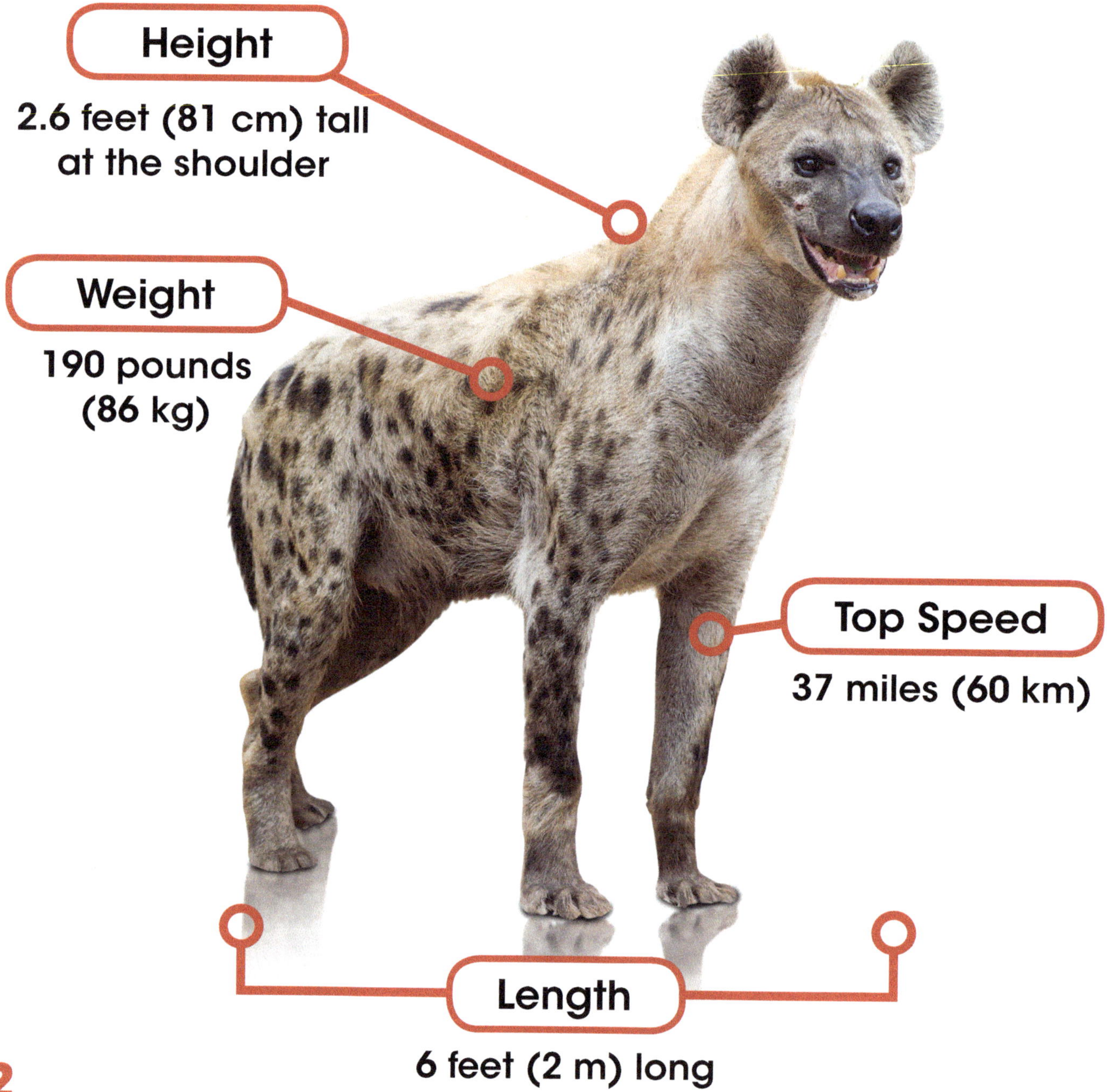

GLOSSARY

communicate: To talk through sounds and motions.

ecosystem: A natural community of living and nonliving things.

grassland: Land on which grass is the main plant life.

habitat: The natural home for plants, animals, and other living things.

scavengers: Animals that eat what they can find, including waste and dead animals.

territory: Living area.

INDEX

WEBSITES

Due to the changing nature of Internet links, PowerKids Press has developed an online list of websites related to the subject of this book. This site is updated regularly. Please use this link to access the list: www.powerkidslinks.com/aotg/hyenas